Funny Bunny Hats

by Ramon Perez
illustrated by Jannie Ho

HMH

Copyright © by Houghton Mifflin Harcourt Publishing Company

All rights reserved. No part of this work may be reproduced or transmitted in any form or by any means, electronic or mechanical, including photocopying or recording, or by any information storage and retrieval system, without the prior written permission of the copyright owner unless such copying is expressly permitted by federal copyright law. Requests for permission to make copies of any part of the work should be submitted through our Permissions website at https://customercare.hmhco.com/contactus/Permissions.html or mailed to Houghton Mifflin Harcourt Publishing Company, Attn: Intellectual Property Licensing, 9400 Southpark Center Loop, Orlando, Florida 32819-8647.

Printed in the U.S.A.

ISBN 978-1-328-77215-2

4 5 6 7 8 9 10 2562 25 24 23 22 21

4500844736   A B C D E F G

If you have received these materials as examination copies free of charge, Houghton Mifflin Harcourt Publishing Company retains title to the materials and they may not be resold. Resale of examination copies is strictly prohibited.

Possession of this publication in print format does not entitle users to convert this publication, or any portion of it, into electronic format.

Bunny makes funny hats.
She made 12 funny hats for Fox.
She made 5 funny hats for Bear.

2   How many funny hats did Bunny make?

Here comes Bear.
"I have 5 funny hats," says Bear.
"I want 9 more hats."
"Coming right up!" says Bunny.

How many funny hats will bear have now?

Here comes Fox.

"I have 12 funny hats," says Fox.

"I want 6 more hats."

"Coming right up!" says Bunny.

How many hats will Fox have now?

Bunny makes 6 hats for Cat.
She makes 14 hats for Dog.

How many hats does Bunny make in all?

Here comes Cat.

"I have 6 hats," says Cat.

"I want 10 more hats."

"Coming right up!" says Bunny.

How many hats will Cat have in all?

It is time to go home.
Bunny puts on 2 funny hats.
There are 15 funny hats on
the counter.

How many funny hats does Bunny have in all?

Responding

Problem Solving

The Cat's Hats

Draw

Look at page 6. Draw the hats Cat has. Draw the hats Bunny has.

Tell About

Draw Conclusions Tell how many hats Cat has on page 6. Tell how many more hats she wants. Tell how many hats Cat will have in all.

Write

Look at page 6. Write how many hats there are in all.